Victor Echo Zero Five

Victor Echo Zero Five

Steven M. Silver

Copyright © 2020 Steven M. Silver
All rights reserved.
ISBN-978-1-7352051-5-1

Cover by Eric Strehl
Blackheart Studios, http://www.ejstrehl.com

Table of Contents

Introduction	1
Fire Flower: Dedication	4
SORCERER'S APPRENTICE	6
Ode to Forts Barancas and San Carlos	12
MAWTUPAC	14
Star Fall	16
Journey #3	18
Winter's Flak	22
Connections	23
Arizona Sky (Unfinished)	25
Grand Canyon	27
FIRE AND RAIN	30
Chu Lai Sand	31
Gone	32
Night Watch	33
Untitled	35
Christmas Past	37
Ajax and a Danang Messhall	39
Flight Line–Night	41
Monsoon Rain	42
Five Minute Alert	44
The Other Breed	47
Close Air Support Mission Number 173	49
The Trembling Earth	50
Down	52
Letter From Home	53
One Aircraft was Lost Over Laos Sunday	55
This Moment	56
Those Old Ban Karai Talkin' Blues	58
Evaluation	61
Three Cons at Two O'Clock	63
Tracer Fire	66
Graveyard	67

Warrior's Lament	69
Laos Escort	71
Flight Deck, Night	73
Stone Cracks (Mission 286)	75
Thunder	76
Dawn Patrol, South China Sea	77
Arena	80
Vietnam Epitaph II	83
Return	84
Victor Echo Zero Five	86
Homecoming	89
SHUTDOWN	91
Yuma Deployment–1971	92
Pompeii's Dust	93
Arizona Night	95
Dry Leaves	97
EC-03	99
Thoughts at 43,000 Feet	101
Vietnam Iliad	105
Mount Out	106
Investigation	107
Tribute: Last Flight	108
OF A FIRE IN THE NIGHT	110
The Wasted	111
Long Gone Time	113
27 Jan 73	115
For Terry Graves	117
Held Fire	119
Those Long Time Rememberin' Homecomin' Memory Blues	120
On the Fall of Pleiku	123
Last Night	127
Late Evening Questions	129
Secrets	131
Brothers and Sisters	133
Of a Fire in the Night	139
Rain	141

Charley/One/Four Remembered 143
Frags 145
Interview–Vietnam Veteran Memorial, 1983 147
On Distance 149
Serenade Echoes 151

Victor Echo Zero Five

Introduction

These poems were written between 1968 and 1990. I did not keep a formal diary or journal but the poems reflect the changes and situations I encountered at the times they were written. They reflect my own thoughts and feelings and should not be seen as a representative statement for all Vietnam veterans. Indeed, when I first began to write it was to put into words the fierce joy I felt in flight; there was no intention to trace one person's journey through a war and its long aftermath. That writing came about because I simply did not trust myself to say to anyone, even those few who wanted to hear, what I felt.

I am pleased that some of my brothers and sisters of the Vietnam War read a few of these poems. They were kind with words of praise; they honored me when they felt stirred into speaking of their own anger, sorrow, and, yes, sometimes, joy.

It is, of course, tempting to rewrite these poems, for the awkward phrase and miss-chosen word now seems to leap off the page and mock my efforts. However, they are as I wrote them, with only spelling, grammar, and punctuation corrected since they were first put on paper. The words are as they were written at the time–as one of my Marine Corps teachers once said, "You owe it to yourself." There it is.

The first poems were written by a very green Marine Second Lieutenant desperately trying to learn the fine art of flight in the McDonnell-Douglas F-4B Phantom II. Some of the later ones were written by a grimmer First Lieutenant caught up in the war. A few were written by a Marine Captain trying to teach others how to fight, kill, and stay alive in that war. A large number were written in the confused anger, sorrow, and resolution of a young man trying to find the way home. The final poems are from a no longer young man who found the way.

Victor Echo Zero Five

No one finds the way back alone. Those who helped me, including my best friend, were my family and some extraordinary friends. And there was my second family, the men and women who also served in Vietnam – when I rediscovered them, a chunk of the loneliness crumbled away. Some of the most beautiful of these folks came to me seeking help with their own dragons. Their strength and courage and caring, even in the midst of awesome darkness, could not help but encourage this all-weather wolf in his own struggle.

They have taught me to never let go.

And then there are the others who rode lightning during those years of the story, sometimes just a few feet in front of me, and more than one has had a chance to look this over and comment – still, any errors of fact are entirely mine.

So I wrote this because I had to, and, because what it was about was as close to being sacred as I have ever experienced, I have tried to tell the truth. All of that is the why.

The what is the story of flight in an F-4 as seen by a Marine in the backseat and what it left me with. I have been encouraged by the stories told by C. F. Rawnsley, Pierre Clostermann, Johnny Johnson, Bert Stiles, and so many others who have been a part of military aviation and had the talent to describe it. I was encouraged by them but did not try to write like them – their styles (and stories) were their own and I thank them for their examples. I can only hope that, whatever far skies they now fly, they would not be terribly displeased with my efforts.

Since poems served as a journal, I usually did not attach any explanatory notes – like most journals, I did not expect to be sharing this. To differentiate between any notes I wrote at the time and those I've added for clarity, I decided to keep things simple by using the term "Notes" for any added information.

Finally, a short bio might help keep things straight as you go through this. I served in the Marines from 1967 to '71, returning in '72 to spend a few months with the Historical Division, picking up a Master's in American History between the two tours. I was a high school teacher for two years, lost my job to a cut back, picked up a Master's in counseling, and worked in a community mental health agency from '74-'78, when I went for my Ph.D. in psychology.

Victor Echo Zero Five

I ran an inpatient PTSD treatment program for 26 years before retiring from the VA. I served in the Pennsylvania Army National Guard as a psychologist from 2006 to 2009. Currently, I am a consultant to psychotherapists.

Victor Echo Zero Five

Fire Flower: Dedication

Fire flower
In my throat
Burns,
Erupted from my soul,
As remembered kerosene
And blood
Mixed with superheated air
And dreams,
Exploded in high thunder,
Echoing in mountains
And clouds.

Above where now I trod,
I once had a place.

Above where now I trod,
I once had a meaning.

Our steely shark screamed,
Ripping white across blue sky,
Which screamed.

Tell me if you have flown,
And I will tell you if you have
Lived.

Electronic crosshairs

Victor Echo Zero Five

Punctuated flight's odes,
Swirling in my heart,
Hunter's heart.

Listen, now, child,
And here you will
find a war-chant
As new as,
As old as,
Man.

Beauty was the cause,
Death was the mission,
So they mingled,
And my once sharp eye
Is hazed with the shadows
Of time,
So I cannot find the thin line
Between the two.

Forgive me not these trespasses,
For if I could,
I would do it yet again,
In a fast ship,
Into harm's way.

And I don't know why.

And I don't know if it matters.

–For VMFA-115, VMFA-531, VMFA-122.
For George McGaughey, Skip Sharp, John Fogg.
For those who waited, and for all those they waited for.
And especially for those who still wait.
And for those still awaited.

Victor Echo Zero Five

SORCERER'S APPRENTICE
1967-1969

Note: Thanks to a publicity machine second in success only to its combat ability, the United States Marine Corps has taught anyone not out of reach of modern communications that, "Once a Marine, always a Marine." There is a great deal of truth to this comment. What the Marine Corps does to its members is help them discover that they are able to do far more than they ever imagined.

I sought membership in the Corps for several reasons but that was one of them. I believed that I had had a sheltered life and had done little to learn about leadership and responsibility. The Marines, I was convinced, could teach me those things – in effect, help me grow up.

Of course the same lessons might have been learned in any of a number of other ways. In 1965 I seriously considered the Peace Corps as an option. I was a child of the New Frontier and spent serious hours discussing the possible assignments the Peace Corps offered with a friend who did join them and found himself in Africa digging clean water wells.

But there was another reason I looked toward military service. The son of a career Navy officer, I had seen some of the world and, though still young, ignorant, naïve, and all the other virtues of youth, I had come to appreciate what my country had given to me.

This made me a very strange bird in the flock of my generation, many of whom saw as their primary duty the engagement in as many self-centered activities as possible. Those who believed they had some kind of obligation to others seemed in a minority and those who thought that obligation was to their country were in a minority yet again.

Just to make it more interesting, I was 4-F. I was not going to be prey for a Selective Service Board and I knew it. "Lazy eyes" and poor vision meant, then, that if I was going to serve in the military, I would have to step forward and ask for it.

I have discovered that being told I cannot do something, that I was incapable of doing something, pretty much gets me to do it. Such are the weaknesses of character that drive us onward.

All right, the military. But why the Marines?

Victor Echo Zero Five

Captain Bob Neff supplied the answer to that. He signed me up for the Marine Platoon Leaders Class while I was in college. A recruiter of officer candidates, his "multimedia presentation" was a three-ring binder that had pictures of PLC candidates going through bootcamp in Quantico. One page in the binder had the words, "Is it hard?" When you flipped it over, the next page read, simply, "Yes."

So there it was.

Of course, there was more to it than that. As a "Navy brat," I had been around Marines my whole life. They were not just people in blue marching by in a parade to me. They were everywhere, doing whatever they were supposed to do with a quiet, competent professionalism, sometimes marked by a wry smile. They took what they did seriously and I always felt, whether I was watching them manning the front gate of a naval air station or working on the mystifying internal organs of an ancient F-9F Cougar, they would get the job done.

I wanted to be that competent in life, as many young men do. In the course of our travels as a Navy family, I had been in eight elementary schools, two junior highs, and two senior highs – the lack of continuity in any one place had left me, on the negative side, feeling a bit like the odd man out, always having to work to fit in and then moving on about the time I did. College – two undergraduate schools – didn't help much until I befriended two young men shortly after deciding to join the Marines.

Terry Graves and Rich Higgins were Navy ROTC midshipmen and both intended to take their commissions in the Corps. Their friendship during my last two years of college served as an anchor as did my membership in the Marines Platoon Leaders Class. Terry died in Vietnam, earning a Medal of Honor; Rich was murdered by terrorists in Lebanon while serving with U.N. peacekeepers.

I joined the Marines with a good idea of what PLC bootcamp was going to be like. My idea was an underestimation of what would be required to make it through. Some numbers might help in understanding what I mean. In my first summer at Quantico, I was in a platoon of 55; 17 would eventually be commissioned after the second summer. Most finished the first summer but many decided not to return for the second six-week trial.

That attrition was not exceptional. One morning during my second summer, as I stood as "road guard" for my platoon crossing a highway to get to the area used for PT, I watched seven shaven-headed officer candidates double-timing behind their platoon guidon. Odd – were they some special unit, some detachment from their platoon?

My Drill Sergeant appeared next to me, ghost-like, his lips inches from my right ear, and he chastised me for not keeping my eyes on the road I was guarding. He did this without repeating himself for several minutes which felt perfunctory and perhaps was, as we were very close to finishing our second summer. Then, once assured that my eyes were locked on the crossroads with an intensity most radar would have admired, he leaned back.

"Do you know who those people are, candidate?" he asked.

"No, Drill Sergeant!" I shouted. There was still some time to go before I would address a Marine at other than attention and with a raised voice.

"That's an OCS platoon, candidate." I could tell he smiled, possibly because I heard his unaccustomed muscles creak as he took on that expression. "They had fifty, now they got seven. And they *still* ain't done. Road guard in!"

I ran back to my position in the platoon and fell into step with the others. Forty-three out of 50 gone? That's a drop-out rate of 86%. How have I managed not to end up leaving?

They made it easy to leave. Just say you quit. There was a point during my second summer of boot camp where I became determined to quit. I hated it, all of it. The lack of sleep, the harassment, the physical exhaustion, the never being good enough.

I just was going to wait until we finally got to the top of the next peak in the Hill Trail – it seemed too easy to quit at the bottom of the hill. Then, once on the top, well, it was downhill from there, so no point in quitting then. And then, at the bottom of the next hill…

Virginia in summer, the humidity of the nearby Potomac adding its touch, at one point collapsed me. We were going up the fifth hill of the Hill Trail. Someone behind me yelled, "Charge!" The candidate behind me said my extending foot never touched the ground and I slammed face-first into the ground, my weight aided by the M-14, steel helmet, and backpack and blanket roll I carried. He tried to grab me as I rolled by. I have no memory of it, nor of finally hitting a tree halfway down the hill and coming to a halt.

My Sergeant Instructor told me on our last day that he tried to take my rifle and I refused to let him have it – it amused him that, even in the delirium of heat exhaustion, I was acting like a Marine. He dumped a canteen of water on me, got me on my feet and pointed me uphill. After gaining the crest, I caught up to my platoon while stumbling into sergeants and officers along the way, a measure of the degree to which I was out of touch with reality.

I caught up to my platoon and would have gone passed them in my daze had someone not steered me to the side of the path and got me on my back. I became aware of my situation as another canteen was dumped on me.

My reaction was embarrassment and then anger. I hated the Hill Trail. Not fear – our relationship was strictly one of hate. Every time the platoon went on another march on it, I was all but snarling at it, determined to defeat those hills.

The hills, of course, were incapable of noticing my emotions. Thousands of Marine officer candidates already had struggled up them, or tried to, and thousands more have since. My anger was, also of course, at myself.

On our last march, we went over the Hill Trail and its adjacent trails – South Bank, Haunted House, and Power Line, names familiar to many Marine officers over most of the past century. And we double-timed much of the route. I remember looking over at our platoon's DI and taking the tremendous liberty of smiling at him. Perhaps because it was our last few days, perhaps because he remembered my fall, and perhaps for reasons I will never know, he smiled back.

All Marine Drill Instructors are secretly teachers and what they teach is how to find out who you really are. Yes, they begin the process of teaching you the skill set that you need to acquire to be effective as a Marine. But the most important teaching is in the form of discovery. The first discovery, of course, is that you can do more than you imagine you can. Then they take what you have found and shape it into a form suitable for a Marine.

The day before graduation we turned in our equipment. Our platoon candidates served as squad and platoon leaders so as to give us practice giving commands and we came back from the supply building under the command of one of our candidates. We crossed the grass of Corsair Field and the candidate gave us the command, "Route step," which meant that we were not going to try to maintain everyone in step in the grass. We could talk among ourselves as we walked and we made small jokes about being almost done.

Ahead of us, in the middle of Corsair Field, stood our Sergeant Instructor. A small man with a red eye, he watched us approach with his hands on his hips, frowning. Clad in green utilities with a silver-colored helmet liner bearing a green stripe and a Marine emblem, his eyes tracked us like lasers. I am certain that in my six weeks with him, with his face sometimes millimeters from mine, I never saw him blink.

Our joking ceased and unconsciously we made sure our dress and cover, our alignment, was correct. Our candidate platoon leader kept us in route step as trying to march in step on grass was not likely to be successful.

As we came alongside our Sergeant Instructor, he growled in a voice that easily carried to the last one of us, "All right, Marines – get in step!" Forty left heels slammed into the ground together. This was the first time anyone had called me a Marine.

We marched to our barracks area like a machine, our soft-soled boots hitting the pavement in perfect unison and with a power that vibrated windows. My father and two brothers, visiting for the graduation, waited in a car nearby and they were astounded by our arrival.

So was I.

On my graduation from college, I was commissioned a Second Lieutenant and returned to Quantico, this time to learn all the things a Marine Lieutenant needed to know.

The Basic School is always referred to as TBS – the inclusion of the first letter is to avoid providing too much mockery. But there was little humor at the time. The Vietnam War was increasing in intensity and three full companies of lieutenants were due to arrive except…

Except there weren't three full companies. Basic School Class 6-67 only had enough people coming out of the Naval Academy, Navy ROTC, PLC, and OCS to field two companies. P Company, to which I'd been assigned, dissolved, forever after known as "Phantom Company," a significant name I did not note at the time. I found myself in the fifth platoon of N Company.

While at Quantico I tied the range record with the M-14 with one round not counted – my instructor later determined it had actually gone through the shredded area of previous hits – and I had a conversation with an officer who suggested that I put in for the infantry with a view towards being the platoon leader for a group of snipers. That seemed like an interesting way to spend a year in Vietnam.

Vietnam had been on the back pages in 1965 when I joined the PLCs but it was very much up front in 1967. We were going to be permitted to make our "wish list" of desired assignments and I thought that the infantry was still my first choice, whether or not I ever got close to snipers, or perhaps tanks would be interesting. Either job promised to be in the heart of action in Vietnam.

We had a meeting with the representatives of various MOSs and I went over to see the captain for aviation. I hoped to snag a couple of posters of Marine aircraft. My brother and I had been devoted model builders and I thought he'd like one.

"Are you interested in flying, lieutenant?" he asked. He was a thin man and was the only representative wearing the Marine white uniform. His gold wings glowed but he wore no ribbons – he was the first Marine aviator I met, but not the last, who did not bother to wear other than his wings.

"I can't, sir," I said, motioning to my face. "I wear glasses."

"Have you ever heard of Naval Flight Officers?"

"Sir?"

"They fly in A-6 Intruders and, of course, in the F-4 Phantom."

"Phantoms?" Fighters? I could be in fighters?

There seemed no other choice. To be good enough to be in fighters, now there was a challenge. On top of that, for there was and is no shortage of challenges in the Marine Corps, there was all I had ever read about fighters and their pilots. Most especially was C. F. Rawnsley's *Night Fighter*, the story of his combat flights with John Cunningham in the RAF. Rawnsely was the fighter's radar operator. He also reloaded the guns. In the F4, the Radar Intercept Officer (RIO) handled the radar, most of the communications, what passed for navigation in the Phantom, helped with weapon release parameters, and supplied another pair of eyes.

From Quantico, then, it was off to Pensacola, where I spent hours bored near catatonia learning how the Navy trained navigators for multi-engine aircraft. When the casualty lists from Vietnam came back, several of us decided to drop out and join our classmates and friends. Captain Dick Morrisey, the Marine RIO representative, talked us out of it, telling us we would do more good in the air.

I decided to stay with it.

Victor Echo Zero Five

Ode to Forts Barancas and San Carlos

From the sea, and low,
The twin-jet trainer came in,
Banshees in her throat,
Flaps and gear down and locked,
And it swept over the beach and,
Flashing shadow on ramparts
Three hundred years old,
Screamed over the masonry
Of the two forts made one.
There are no ghosts here,
Now, in its galleries and gun pits;
They've fled before the near sonic wail,
And only silence and dust remain.
From its heights, where five flags flew,
The beach and bay can be seen and guarded,
Or could have even sixty years ago,
But the deadly machines,
With banshees in their throats,
Made its gun ports a mockery,
Worse, an anachronism.
So out of shame and insignificance,
The ghosts left, leaving the fortress,
Scarred masonry and bloodied walls,
To the inquisitive young eyes and minds
Of two curious warriors in a fort
That died six decades
Before they lived.

Victor Echo Zero Five

Note: The double forts covered the approaches to Pensacola and were first built by the Spanish, occupied by the Americans, occupied by the Confederacy, and then occupied by the United States once more. This was the first poem I wrote when I arrived at Naval Air Station Pensacola for training in the BANAO (Basic Air Navigation and Observer) program. Training became very time consuming and I did not resume writing until I was assigned to VMFA-531 in California.

Victor Echo Zero Five

MAWTUPAC

Share the sky
with hawks,
turn and soar
seven miles high,
then whipping by whitecaps
and...
...the centerline
station is normally
loaded with no more
than three 550 pound
napalm...
Racing the wind,
outrunning our own
sound, opening our
minds' horizons
to God's...
...Mark four gunpod
capable of four thousand
rounds of 20 millimeter
ammunition per minute,
consisting of ball,
armor piercing,
explosive, incendiary...
Straight up,
with chained thunder
pushing us through
head-on intercept

Victor Echo Zero Five

so as to obtain
maximum displacement for
both radar and heat
missile launches...
...And damned be
the man who first put
insignia on a wing
and guns on a
plane.

Note: Marine Air Weapons Unit, Pacific, based at El Toro Marine Corps Air Station, is where I was introduced to the fangs of the Phantom. I was assigned to Marine Fighter Attack Squadron 531 as a "nugget," someone just starting out. VMFA-531 served as the West Coast training squadron for F4 aircrew, pilots (Naval Aviators) and RIOs (Naval Flight Officers). Eventually, there would be specialized training squadrons (VMFATs) but that would not happen until the Vietnam War faded.

Victor Echo Zero Five

Star Fall

I

A line among the stars,
It glowed white and was gone,
With a roaring silence,
A far ranging rock, too eager,
Had touched Terra's breath,
And burned, on Christmas Eve.

II

In military green he saw star fall,
Trained eyes noted it, discarded it,
From behind steel-framed Plexiglas
A momentary comparison was made,
Thoughtlessly, to a warhead's reentry,
(A streak of light followed by ultimate light)
And then he turned the fighter homeward.

III

A man-cub, a boy, watched the stars,
His breath a quick, cloudy ghost,
His hand held, softly, by his mother,
Who said, "Make a wish, it's a falling star."
They turned to go, though his eyes still,
And sadly, reflected the sky's diamonds,

Victor Echo Zero Five

And he said, "I wished that the star,
The dying one, could go back into
Heaven where it belongs."

Journey #3
(For "Buddha" McClennan)

I

As a child, I dreamed
And saw, in clear night,
The stars.
Beyond touch, but just,
Beckoning, not mocking,
The child who dreamed,
And saw, in clear sky,
The hawks,
Soaring proudly
and saying, "This way,"
And the child heard.

II

Childhood's end, but,
Forever, the dream,
The eternal dream,
And so it was I
Found a way to
The skies.
Many learned the way,
And chained thunder
To steel and aluminum
And learned to think
"We."

III

It stood taller than Everest,
Churning, massive, boiling,

Victor Echo Zero Five

White column, gray based,
A monstrous, wildly living
Tree, its roots living lightning,
And it could level forests;
Call it Thunderstorm.
If it could, it would kill us,
Would bend steel
And scatter us in the wind,
Shattered dry leaves.
But, mightiest of the mighty,
It cannot, we are above it,
Higher than any hawk,
Faster than the storm's
Thunder.
But there is never
Any temptation to mock it,
For there will come a time
In our journey
When we cannot avoid
The storm, trapped,
And we will pray
For the storm's mercy
And the competence
Of strangers who
Fashioned our far ranging
Craft.

IV

We cause the sun to rise
And set, childishly,
And rejoice in our freedom.
We are, in one moment,
A silver glint, arrowhead,
Leading thin contrail
In glowing sky, and,

Victor Echo Zero Five

We are, next moment,
Below the tree tops,
As twenty-five tons of
Shark-shaped metal
Scream into harm's way,
With hell on board.
But mostly, we find,
Quietly, the beauty
Of the high journey.

V

Our high journey
Is never done alone.
There is always
Gann's eternal hunter,
Called fate, who
Reaches out
And the land
Is scarred and burned
Where we fell,
For there is little room
For error, yet we
Are still men,
And we do err,
And sometimes
We do not err,
Yet fate is still
The Hunter
And the night sky
Is ripped by
A waterfall of
Flame and
Screaming metal
And unanswered,
Ever, is

Victor Echo Zero Five

"Why?"

VI

I am young, still;
The high journey
Has not yet
Demanded the partial
payment of my youth
Nor the Final Payment.
And I know the
Cost, in some amount,
Must be paid, yet
I will go on
Across the Sun's face,
For a while longer,
If allowed by fate,
For I love
The high journey.

Victor Echo Zero Five

Winter's Flak

Leaves spin in,
Hit by Winter's
Flak.
An airburst of
Cold cut
Chlorophyllic
Hydraulic lines,
Snapped
Cellular
Structured frames.
Deprived of
Aerodynamic
Support
The drag and
Weight were
Greater than
Lift and thrust.
So it stalled,
Spinned,
Crashed,
Burned,
And died.
Always said
He don't
Make'm like
He used to.

Victor Echo Zero Five

Connections

Oxygen, from steel ball,
Through tubing, steel and
Rubber,
Through valves, steel and
Flesh,
Into blood.
Electrical impulse
To muscular cells
Expansion,
Leverage to
Cables and pulleys,
Hydraulic lines
To flaps
And rudder
With feedback.
Ultra-high frequency
Vibrations
Are selected
And converted
To electrical
Modulation,
Causing low frequency
Vibrations in earphone
And air
And ear
And is reconverted
To electrical

Victor Echo Zero Five

Modulation.
They are blended,
They are one.

Victor Echo Zero Five

Arizona Sky (Unfinished)

We came here, he and I
And she,
To fashion our skills
At the scientific art of
War.
In the false water glistening
On the runway, in our fighter,
Nestled in her body,
We come to hate waiting
And the sun of the Mescalero,
As our life juices stream down
Our enclosed bodies.
She hates the terrible heat,
Makes her sluggish, like a tired
Rattlesnake, then...
Then...rebirth.
Throttles slam home,
A growl is a roar,
Shaking souls miles away,
And twin pillars of clear flame
Lance out, and we are no longer
Earthbound.
Newly aloft, full fuel, bombs,
We are heavy and oddly,
At three hundred and eighty knots,
We are still slow.
Here, in Arizona's sky,

Victor Echo Zero Five

We look below and wonder
At those who came this way,
Those many years ago.

Note: This poem was begun in 1969 while deployed to Yuma for bombing practice. For one reason or another, I did not finish it. When I returned to Yuma in 1971 after a year of war I tried to finish it–it seemed very important to finish it. My concentration was disrupted by my aircraft's attempt at self-destruction on, or rather – and most abruptly – off, the Yuma runway. Now I am too far away from that most brilliant of skies to finish it.

Victor Echo Zero Five

Grand Canyon

Arizona desert
Blurs past.
Dust devils and
Even hawks
Stand still as
We pass.
Low, over the Colorado,
We twist and turn,
Following the river's jinks,
Then flash across lakes,
Startling early morning
Fishermen
With our whistling thunder.
Within canyon walls,
Illegally, we race still,
Following the river,
Testing our skill and
Our craft
As brilliant, colored
Rock whips by
Our wingtips.
Ahead, suddenly and
Too close,
A silver necklace
Across our path.
No time to do else
But see;

Victor Echo Zero Five

It is ahead,
It is above,
It is behind.
Clear, we climb into
The sky's freedom
And spiral upward,
On our backs,
A victory roll,
For we have won,
We reached out
And pulled
The Old One's beard,
And in the foolishness
And bravado of our youth,
Our challenging youth,
We rejoice in being alive,
Even as we arc and loop,
And reenter the canyon
And follow its contours
And twists, until again
We race under the power line,
Then run fast and free until
Wide lake reflects
Our gray and white image.
We curve up and turn,
Knowing our thunder will echo
Back through the canyon;
We salute it, and ourselves,
For having the spirit
To dare ourselves
And the sky.
Turning homeward,
We reflect, older now,
Years in minutes,
And understand a little
More of the sky.

Victor Echo Zero Five

It is done, we will not, need not,
Challenge the canyon again.
Now we know.
Thank you, Grand Canyon.

Victor Echo Zero Five

FIRE AND RAIN
1969-1970

Note: It had been the custom, I was told, for Marine squadrons to rotate between Japan and Vietnam. Then the decision was made to just rotate personnel. I arrived in Okinawa and was offered the choice between going to Japan or Vietnam first. Getting the nasty stuff out of the way seemed like a good idea so I said I wanted Vietnam. That same week the policy changed again. Now aircrew would go to either Vietnam or Japan but no more switching. It seemed a coincidence that the same people I knew in my F4 squadron in California, VMFA-531, now showed up in the same F4 squadron in Vietnam, Marine Fighter Attack Squadron 115. It was not a coincidence; a network of old hands ensured that most of the "shopping list" of one of our 531 instructors was used by a buddy for assigning us together. When the F4 squadrons were withdrawn from Chu Lai, I was transferred to VMFA-122 and went with them to Hawaii, cutting my Marine-stipulated tour short by two weeks.

Our missions were of several types. "Close air support" assisted troops in the field and included blowing up tree lines that contained bunkers and trenches but not necessarily any enemy soldiers. Sometimes it meant clearing the trees off a hill that might, or might not, be used later for a helicopter landing zone. And sometimes it meant striking while our troops were in hand-to-hand situations. We also struck the road network in Laos, escorted A6 Intruders at night as they sought to hit trucks, and covered recon aircraft following the Ho Chi Minh Trail. We were tapped to provide combat air patrol for the Navy off the coast of North Vietnam and B52s working on the Trail.

Victor Echo Zero Five

Chu Lai Sand

It slips past
Windborne,
slow tendrils
ripple at your feet
and are gone.
Uncaring, unseeing,
all penetrating,
drifting its seed
into all
sanctuaries.
Flowing, alive,
it rolls around
and over all,
with a quiet
hissing whisper.
Pauses and rises,
exposing strength
in muscled dunes
that in the night
vanish.
From nowhere
to somewhere,
unfettered
comes the
eternal sand.

Victor Echo Zero Five

Gone

The fighter tumbled
into the floor of clouds;
"We're okay," Peck said,
Then vanished.
Gray monsoon sea
blundered beneath us,
merged and swirled
with drifting scud
as we searched.
"Peck, are you there?"
Nothing–mocking sand,
mocking sea; no,
not mocking, worse,
indifferent.
No scattered metal,
no blackened earth,
no rainbow slick on the sea,
nothing marked their fall,
from sight to only mind,
dimly seen.

Victor Echo Zero Five

Night Watch

The dark quiet
is lightly broken
by the gentle
monsoon rain
and is torn
by the hunting
roar of the lean
gray phantoms.
But, mostly, there is
just the ant-steps
of the patient rain
on the aluminum roof.
He sits, slid
back in his chair,
clad in green,
a flight suit
that will never
be free of
his smell.
His companions are two
telephones, usually silent,
but possessing the
potential of unleashing
man's hell on other
men.
The book he tries
to read,

Victor Echo Zero Five

which he's read
before, is old and
worn and fails
to merit even as
good pornography,
but there is
nothing else in the
night until the last
wailing fighter comes home,
then he can sleep,
and in this night watch
the thought is
good.

Untitled

The madness,
blood-red and black,
caught hold
like a wildfire
in the wheat field
of a man's soul,
and then was
My Lai.
My Lai, dear God!
My Lai!
Speak not to me of Hue and the 3,000,
of the work
of the foe;
my ears were
deafened by the
cry of a child at
My Lai.
Speak not to me
of war's bitter
fortunes, for I,
a warrior, am the son
of a warrior
and know where war's
awful lines must end;
at My Lai.
The children,
whose tomorrows
will never be,
whose tomorrows
we fought to defend,
we bled and died for,
they are gone.
He betrayed us,

Victor Echo Zero Five

mocked our cause
and stained our honor
with the blood
of children
from My Lai.
I will stay
for our cause and
for our duty,
but the outrage
of My Lai,
and the betrayal,
will be a cold
rock lodged deep
in what is my
heart.

Note: I learned of My Lai in bits and pieces during 1969 and wrote this on Dec. 9, 1969. I can't speak of others' experiences; it seemed to me that everyone I encountered worked hard to avoid harming innocent people. But war is organized savagery and a dropped bomb goes where the laws of physics dictate, regardless of the efforts of the aircrew. Only once did I think we might be directed to put our bombs on civilians. They were using a road adjacent to a patch of woods in the middle of which a South Vietnamese Army unit was in a firefight with Viet Cong and our Forward Air Controller, the person directing the strike while in a small, twin-engine aircraft, seemed to be taking a long time decided where he wanted us to drop our bombs. I decided if, for some reason, he was going to call for our bombs to hit the road, they would never leave our Phantom. Beside my right calf was a panel of circuit breakers. I counted down a row of them to J10, the breaker that controlled the drop signal to our bombs, and got ready to pull it. And maybe face a court-martial. But he took time because he wanted to make sure he could get our bombs in the right place to help the Vietnamese army while not harming the people on the road. He did and we did.

Victor Echo Zero Five

Christmas Past

We lay in the straps and buckles of flight gear –
unsummoned gladiators in a hot-pad van –
trying to disappear into sleep's wilderness,
when sharp, jolting buzzer jerks our nerves
and our bodies into sudden rehearsed movement.
Quick run across rain-pooled concrete to Phantoms,
crouching within steel revetments, monsoon overcast
shrouding all in dull gray light,
light enough to see by, to merge ourselves,
wires, buckles, hoses, body to body,
we to Phantom.
A quick wipe of flight suit sleeve across
rain-smeared forehead, sweatband pulled on,
black and green helmet pulled tight
to hear wingman call up.
Unnaturally calm voice into oxygen mask
as I breathe her, the Phantom's, cool breath,
and we are moving through the mist,
impatient, even in our fear, to gain the sky,
to be hunting yet again.
Sixty-three minutes later we ease over the gray sides
of our now still fighters, etched lines
in our faces from masks and around our eyes
from fatigue of more than body.
Slow walk back to the 'pad, shoulders sagging,
and as we enter, someone looks up to say
"Merry Christmas. They already called in your

Victor Echo Zero Five

body count. Fifteen
crispy critters."
We shrug and drop onto cots – third
mission since midnight leaves little but ash.
Skip looks over with red-rimmed eyes, Texas drawls,
"Fuck it. Don't mean nothin'." He's right,
Chu Lai Christmas, 1969, don't mean nothin',
Not a thing, not today, not yet.

Note: Bob Hope was at Chu Lai putting on a show when we launched. I was back in the world when I saw a retrospective of his many shows for the troops over the years and he remarked, while showing his Chu Lai '69 show, on the sound of jets in the background. That was us. This poem was written after we got back to the hot pad, a trailer with old cots and a field telephone for scrambling air crews in response to emergency situations.

Victor Echo Zero Five

Ajax and a Danang Messhall

We sit, now, grinning, quiet,
in a Danang messhall,
Skip – Ajax Lead Alpha – and
I – Ajax Lead Bravo – while
our Phantom – Victor Echo Zero Five – finds
refuge and repairing kindness
at the hands of strangers forty-nine miles
from home, silent smiles knowing
it could be further,
delighted it was not,
sobered in understanding not by us
was the difference made.

Sugarbear walks on by,
I wave and he wonders,
at my presence here,
as I do, but we
wonder for different
reasons.

Two hours before...
blinded, lost,
only impassive undercast
met our searching gaze.
Voices weakly came,
as we called for help,
but we and our sick bird

Victor Echo Zero Five

could not be found
until Charger 4-2 said
"Here!", and they
on the ground led us
through monsoon's
mist, past clawing hills,
to a Danang messhall.

So we sit, enjoying
plain food and
enjoying being able
to enjoy.

I wonder on the why,
but find no answers
later in a squadron bar,
so I raise a coke
to Ajax, Victor Echo Five,
myself, and
Charger Four Two,
and listen to the laughter
of the gods of the
South China Sea.

Note: Skip Sharp was a pilot who wore his airplane and easily was one of the finest I flew with. We had lost the use of our transponder that ground radar sites needed to see us. Our radar was out, our navigation gear had failed, and we were down to using one frequency to talk to Danang as we tried to figure out how to land on a runway surrounded by mountains at night and in the rain as our fuel ran out. We had decided that we would eject, if it came to that, over the ocean to keep our plane from falling on the city of Danang. But Charger 4-2 got a direction cut on our radio transmissions and, coupled with Danang's cut, they could pick out our weak return on their scopes. Once we landed and were repaired, we flew down to Chu Lai. I was not ready to sleep, so I wrote this.

Victor Echo Zero Five

Flight Line–Night

A careless wind in the night,
off the South China Sea,
slipped between the dark hangars
and caressed the flight line.

Starlight and moonlight shine coldly
over the black shapes
and reflect streaks on formed steel
and crystal canopies.

No, defiant thunder stilled,
no sounds disturb the sleep
of these man-made gray Phantoms,
save the tick of steel in the wind.

Truly alive, they sleep now,
as to the men who fly them,
strike far, strike fast, strike first,
a credo, is stilled this night.

Victor Echo Zero Five

Monsoon Rain

Monsoon rain
drifts across
Chu Lai sand.
Haze gray sky
melds to the sea.
Dark scud overhead
slides on hesitant breeze
through misty rain.
Dark green hills
are lost in swirling gray,
but echo a questioning,
muffled, whistling roar.
We stop our walk
and silently search above
as the invisible Phantom
howls by.
Momentary silence,
then the growling comes again,
in from the south,
and we strain to see.
Suddenly it is here,
racing across threshold lights,
reaching for the ground,
nose pitched up and eager.
Wheels touch, rain swirls,
A chute blossoms,
and hook grabs cable

Victor Echo Zero Five

and they stop between
yellow arrows.
We turn and walk,
sloshing through water,
feeling what they feel,
and curse monsoon rain.

Victor Echo Zero Five

Five Minute Alert

Beneath Asian sun
gray shapes wait,
angles punctuated by
red or green lights,
umbilicals to
wheeled yellow boxes
hang heavy from
fighters' bellies,
tethering us to earth.
We await a word,
strapped and wired
and blended in
our cockpits,
as the patient
sun crawls
overhead.
The necessary numbers
are ready:
mils,
frequencies,
settings,
fuses,
channels,
quantities;
the crossbow is
cocked.
We sit,

Victor Echo Zero Five

studied nonchalance,
and wait.

It is the first hour.

All is done;
our movement
ceases.
Only the
sweep hand of
the clock gives
any indication
of change,
while around us,
in t-shirts,
young statues
blink back sweat
and wait.

It is the second hour.

North, a hundred miles,
there may be a fight,
we wait,
a crossbow bolt,
for a chance,
for a resolution.

It is the third hour.

From the corner of
my sweat burned eye
I see him come, trotting,
not hurrying in the
heat, and I know.
He approaches,

Victor Echo Zero Five

passes his hand
in front of his
throat,
and the power units
whine
down.

Victor Echo Zero Five

The Other Breed

("There are two kinds of people in the world–those who fly, and everybody else.")

They walk slowly
down a dusty road,
different, yet in
sweat stained green
and pulled low baseball
caps, they are much alike.

(A half mile away
a hand moves forward,
kerosene sprays into
superheated air, explodes,
shock wave diamonds
form in columns of flame
and twenty-five tons of shark-gray
metal begins to strike out,
wailing her banshee
song.)

The rumble comes,
quietly, at first, and,
seemingly, they ignore it,
but they listen,
and the defiant song
comes shaking the
indifferent Asian sand

Victor Echo Zero Five

and they stop and turn,
silent, and seek it out,
this more than way of life,
and, low over bunkers
and weather warped hovels,
they see her.

(Lean and gray,
climbing into her element,
challenging the sky
and all who would hinder her,
clear flame in her tail,
men in her body,
a star on her flank,
and hell under her wings.)

They watch,
impassive,
uninterested,
but,
they watch,
'til it is
gone
and they turn
as their hearts
return.

Victor Echo Zero Five

Close Air Support Mission Number 173

From without:
Crystals of ice
 and fire
light eyes,
 reflected souls,
measure, calculate
 behind plastic,
evaluate, aim
 and guide
hell's lash.

From within:
Eyes burn
 with sweat,
bile tasting sour
 as tracers climb,
the world rolls,
 a stolen voice marks,
red floods vision,
 body sags heavily,
escape is made.

Victor Echo Zero Five

The Trembling Earth

Machines.
Sleek, gray shapes,
half seen, unknown,
as if superstitions,
nebulous and fearful,
marked by dragon's breath,
banshees' wail,
hell's thunder,
and the earth trembles.

Machines.
Split the night sky,
dark arrowheads,
hurtling up and out,
above oriental sea,
through torn clouds,
a passage unseen,
but hear the coming rumble!,
and the earth trembles.

Machines.
Bringing from Apocalypse
the black horseman,
with tearing blast
and clinging flame,
the awful burning scythe,
ripping down from the night,

Victor Echo Zero Five

leave the earth wasted,
and men tremble.

Victor Echo Zero Five

Down

The bird fell, no sparrow.
The fire-flower blossomed
and became a wavering
black weed.
"Goose, are you okay?"
a voice inquired.
It went unanswered.
The smoke rolled away,
caught by the wind,
and on the wind,
"Goose, are you okay?"
Silence said all,
silence.
We waited.
Time was solid.
I touched it,
felt it harden,
as we waited.
The dark silence
erupted in
yellow-white sun
brilliance.
A voice, elated,
"We got Goose
and Johnny Cool–
they're okay."
Time fragmented, melted, flowed.

Victor Echo Zero Five

Letter From Home

Dread and expectancy;
with these we await
the daily mail.
A touch of home,
joy,
but sometimes
horror.

The sergeant's wife
mailed a letter
with Polaroid's
of herself
and another.
Stark flesh and
flesh, intertwined,
mocking.

The sergeant took
his .45 caliber
automatic and
introduced a slug
at 860 feet per second
into his already
shattered brain.
His lieutenant
found the pictures,
"Personal Effects",

and fought tears.

Note: It was customary after a death to go through the Marine's personal possessions and remove anything that might be embarrassing for the Marine or to the family. In this case, I was told, the wife had been trying to get a divorce and so sent the sergeant the Polaroids. In the package sent to her, the lieutenant included the brain and blood spattered pictures.

Victor Echo Zero Five

One Aircraft was Lost Over Laos Sunday

"One aircraft was lost over Laos Sunday."
AP and UPI had their say.
One aircraft,
made in Missouri,
delicate and strong,
holder of the song
of whistling thunder,
was raped by a piece
of metal
made in Czechoslovakia,
and, raped, it bled
and burned;
yet, hurt in Sunday's sky,
struggled a few miles
from the craters of
Tchepone.
"The crew was rescued,"
AP and UPI added.
Saved from their terror
and anger
by strangers and
an aircraft
lost over Laos
Sunday.

Victor Echo Zero Five

This Moment

For this moment
there is no war,
there is no destruction,
just the graceful beauty
of our soaring birds
as they fly with the
exhilaration of youth,
youth now innocent again,
ten minutes out of the Ashau.

We curve upward,
arcing across the sun,
slipping through cloud islands
above the metallic sea,
one moment three miles high,
this moment scattering sea birds,
flashing gray and white
reflections on South China Sea.

We climb, we roll, and
in the incongruity of looking
up to see down,
I see the fisherman,
in a small craft, nets strung out,
and I wonder of his thoughts
as he watches the soaring Yankees.

Victor Echo Zero Five

Does he think we prepare to kill him,
does fear touch him, or
does he think we practice at war?
Perhaps; he has seen us at our trade,
but, as we flash upward clean and free,
does he see us as we are,
this moment,
and does he share in our joy?
I hope it is so for him,
so there is no war,
this moment.

Victor Echo Zero Five

Those Old Ban Karai Talkin' Blues

"They ain't got no name, lieutenant, they's just the blues," said the young man from Detroit, picking gently at his guitar in the early morning dark. "My daddy calls them talkin' blues 'cause you just talk it, whatever you been carryin', you just walk it out with the music." So the lieutenant, still wearing flight gear, sat beside the lance corporal and found, by the time the sun came up, you can walk it out with the music.

Well, I'm sittin' and waitin',
in the shadow of an old,
grandmother old, achin' bone old,
graveyard dodgin', angry lookin',
yeah, an old Fox Four Bravo,
and I mean a Bravo, cause there ain't,
I mean you just can't find,
just never seem to be,
maybe there never were,
cause there just ain't no
Juliets, no, no Juliets,
just old Bravos.

And I sit here, just contemplatin',
the letters I got just waitin',
for me to read, but no time,
just no time,
I got no time at all, 'cause
I've got those old blues,
those old Ban Karai talkin' blues,

Victor Echo Zero Five

I got those blues again, momma,
and I just want to curl up and blow away.

And hey, momma, ask now, what do you say?
That old China wind keeps tuggin' at me,
tryin' to drag me, drag me away,
far damn away from these old
Ban Karai blues,
hear me talkin', those old
Ban Karai blues,
but hear me say, hear me now,
'bout why I got these
Ban Karai blues.

Well, now, momma, there's a place,
Oh, momma, what a place,
so you follow the Ashau,
bob and weave on past Tiger,
kind'a slip around, you gotta step around,
the "Z", and keep goin' north,
gotta go north, go on north,
and find out why I got those
old Ban Karai talkin' blues.

Listen, momma, it gets motherin' dark,
bleedin' dark, about two in the AM,
but that's the time,
seems to be the only time,
and you can run out of time,
when you visit old Ban Karai,
and begin to get those
ever lovin', tough hangin',
Ban Karai blues.

Man, they all come out to meet you,
a regular party, I just wish,

Victor Echo Zero Five

I sure wish, I gotta say,
I wish I wasn't invited,
'cause they don't seem to like me,
no, momma, they're all hard,
and don't have a good word to say,
not ever, not even, not one.

And they're all there,
and, man, they're all there together,
while we bob and weave,
tryin' to fake it one more day,
while Charley makes it 4 July,
and it's still November,
and Christmas is forever away,
and gives those dirty old, those damn old,
I mean you begin to get again,
those old Ban Karai talkin' blues.

You go on home, limpin' on home,
back on a wing and a curse,
slippin' on back in the dark,
tryin' to sleep as the sun goes up,
and then, momma, can you believe, momma?,
the man comes by and says one more time,
do it again tonight, and momma,
that's when for cryin' sure,
that's when you really get,
those ol' Ban Karai blues.

Note: Tchepone, once a village, was at the crossroads of the French road network that was the basis of the Ho Chi Minh Trail. Located in Laos, Tchepone connected to the North Vietnam passes at Nape, Muy Gia, and Ban Karai. The anti-aircraft fire in these places could be "heavy to ridiculous." There were two types of F4s flown by the Marines at this time – the F4-Bs and –Js. The Js had what was supposed to be a superior radar and were stationed in Japan while the older Bs all were in Vietnam.

Victor Echo Zero Five

Evaluation

I have felt the steel I ride
(married to it,
a part of it,
living it)
tremble with a hunter's eagerness
as it found the foe's scent.

A Valkyrie's battle cry
whipped into the wind
of the South China sea,
echoed unheard between moon
and Hainan Island.

The echoing wail
gave way to the rolling thunder
and twin columns of living flame
holding burning yellow diamonds
as we lanced into the night.

I am watched by blind stars
and indifferent sun
as I tumble through a foreign sky
and fight to live
and kill.

Friends fall like autumn's leaves,
the tree lives still.

Victor Echo Zero Five

A path of demon thunder,
we follow it,
rejoice on the little
glory of the road,
ignoring the contrail's toll.

Technological art,
aerodynamic mysticism,
Merlin knew no Phantoms
such as ours.
Dragons and warlocks
fade before the reality
and the dream of
screaming steel.

For I,
the old gods are dead,
slain by the radiated broadsword
beaten from the Wrights' plowshare.

Who now can I stand with?

Victor Echo Zero Five

Three Cons at Two O'Clock

Three cons at
two o'clock
high.
We are curious,
with time,
and, what the hell,
turn and climb
in a gentle arc.

A mile in trail,
we coast just
under the wide,
white contrail left
by the number three
B-52.

It is like flying
upside down over
the snow-ground as
the billowing cloud
path whips past
our heads.

We close, slowly,
and pass
to one side,
staring at

Victor Echo Zero Five

the huge
black and green
shape with the
Orca tail.

We cannot see any men.

The hulk shows nothing,
impassive,
driving its bluntly
streamlined body
indifferently through
the south Asian sky.

There is power here,
not finesse,
in eight engines,
and slung on great
pylons, a hundred
remoras show more
power.

There is solidity here,
immovability of purpose,
but I see no heart;
this is a knight's helmet
with no gleam of eyes
in the visor's dark slots.

A small chill, and we
prepare to leave.

As our wing rises,
I see, within a canopy
fairing beneath the
great black rudder,

Victor Echo Zero Five

a man's face,
a raised hand,
and the juggernaut
becomes more
than just a machine.

Still a weapon,
it now has a paradoxical
pulse of life
within it,
and as we roll
towards the earth
six miles below,
I wonder on the blending of men
and their machines.

How much of which
becomes the other?

Victor Echo Zero Five

Tracer Fire

In the dark sky,
flying black in midnight blue,
an old cargo plane is now
a hunter, circling, casting, waiting,
waiting...
A handful of green dots climb upward,
wide of their mark,
thrown perhaps in fear, frustration,
or even defiance–it
does not matter.
A wing is dipped, no salute,
and flame slashes out.
A hundred, a thousand streaks,
searing red, merge and are
part of a stream clawing,
tearing, at the dimly seen earth.
Beautiful terror, these tracers,
with lives measured in meters
and seconds, with lives ending
quickly,
doing the same to other lives.

Victor Echo Zero Five

Graveyard

North wind, sea born and quiet,
slips over the beach,
across perimeter barbed wire,
whose tangled thorns reflects
white Asian moon, and
tugs easily at the wind sock
and spins the vanes on the
light-jeweled control tower.

It swirls around waiting fighters,
impelled by their hellish breaths,
skirts a silent, deserted bunker,
and enters the graveyard.

Here in the dark lay the hulks,
man-fashioned of exotic alloys,
abandoned, now, and dead.

The men who rode their thunder
gone, all gone, but few died,
for their crafts' last gift
was, for many, life.

To the unknowing eye, nothing,
save skeletal frames,
faded and streaked paint,
sometimes blackly smeared,

Victor Echo Zero Five

twisted, buckled, and torn,
nothing, to the unknowing eye.

But to those a part
of the silver glint trailing
the sky-splitting contrail,
to those who learned "we",
this is a place of strange reverence,
of quiet voices, for oddly,
these remains had pulsed with life
as surely as a high-journeyed hawk.

But now, just the wind moved,
and in the night, old metal creaked,
and remembered.

Victor Echo Zero Five

Warrior's Lament

The night comes,
edging in from the sea,
and the cold quiet
of alone
descends.

The vanishing sun
saw the utter darkness
of the hell on earth
and on men
called war.

In brilliant Asian sky
men died, sweat
mixing with blood,
but just now
comes night.

An emptiness follows,
as night follows day,
the hunter's lust and
the hunted's fear
are gone.

Driven to the fringes
of a warrior's humanity,
we try to regain

Victor Echo Zero Five

what we lost
in war.

But what is there to grasp?
A warrior is forever alone,
the ones we love beyond touch,
so we wait for the dawn
alone.

Victor Echo Zero Five

Laos Escort

At six miles high
we are alone in the night,
bound for a point in the black
determined by a nebulous beam,
to meet and guard another,
who will prowl in the darkness
and slip over the roads,
challenging the trucks.

But the rendezvous is still
many minutes away, so we
are relaxed, listening to others,
and counting the stars.

A voice, and we find our
partner. He is slower,
but he is the bomber,
and he will do what needs be
done. We give him our name,
"Sundance," and he leads on.

On the ground, something stirs,
invisible, yet detected;
two trucks, driven with skill
and courage and luck, strive
for a few more miles.
He has them, he strikes,

Victor Echo Zero Five

and the guns claw at him.

"Sundance in on the guns,"
we say, and they turn on us.
A dozen flash bulbs explode
and strings of red Christmas tree
lights wander around us.
Our weapons strike from our
falcon's stoop, a swath
of sparkling diamonds
on the dark earth and
the guns cease, prudent,
if not dead.

We climb and turn,
"Heading one two zero,"
A common phrase, and we
go home.

Victor Echo Zero Five

Flight Deck, Night

(For Yankee/Dixie Stations)

A world of its own,
enveloped, carried and covered,
by phosphorous crashing sea,
and diamond touched night sky.

The sea, thrust past by man,
and the scattered stars,
are witnesses to the beauty,
fascinatingly so!, of war,
while, on steel plates,
steam whips sternward,
shrouding sharp lights
in glowing clouds,
as catlike men hold a ballet
with many-tonned falcons.

Twin massive arms of fire
slam out, a banshee's shriek
and a god's thunder, and
the miracle of Kittyhawk
touched by eternal Mars,
screams out and away.

And in the night,
the ripple moves of men,
and the machines of men,

Victor Echo Zero Five

are so swift and sure
that the fine line between
steel and flesh is oft',
seemingly, forgotten;
but it is still there,
and in this ballet of
red lights and steam,
and flame and dragon's
howl, a move poorly made,
and the flight deck thunder
is cut with the lightning
of a man's scream.

Two per minute, and more
scream skyward, trailing
solid orange flame.

Then they are gone–gray ghosts
in the night–phantoms,
and the lights are dead,
and the only sound across
the steel plates is the wind's.

Note: The Navy carriers operating in the South China Sea used two stations; off the South Vietnamese coast was Dixie, off the North Vietnamese coast was Yankee. I never flew off (or, much more exciting, onto) a carrier. Consider this artistic license.

Victor Echo Zero Five

Stone Cracks (Mission 286)

In the whipping heat
green growth
writhes in blackening
agony,
curls, nods, falls.
Red boiling blister
rolls upward,
collapses,
fades,
leaves an oily
smear in
Asian sky.
The thunder fades,
pauses,
returns.

"Two's in hot–
Where do you want it?"

Victor Echo Zero Five

Thunder

Thundering, they come,
thundering, not like drums,
whipping over the elephant grass,
and a tearing, rasping, wailing roar
shakes the tree tops.

In this moment,
fear is gone
and from the seared
sky
they claw at the ground,
tear it and burn it.
Thin lines of flame arc up
but the warrior's madness
cannot be denied
and they attack the very
flame.

Thunder in the sky
was, for this moment, matched
by thunder in the blood.
Perhaps later, perhaps never,
will they look back
and understand
the thunder.

Victor Echo Zero Five

Dawn Patrol, South China Sea

Listening to gentle whisper
of man-hurried wind
over canopy and fuselage,
I watch the gold towers
of dawn thunderstorms
slowly slide across
a still sleeping Earth.

Our high thunder
is lost amongst the
flashing growls
of the flickering anger
of the sky.

Under one wing, the sea,
under one wing, the land;
all about, the sky–
an awesome familiarity
is building within
as a home is found
three miles above
the world's solid mass.

Morning star flickers
a coded message read
by my heart, a challenge,
"Who goes there?"

Victor Echo Zero Five

The reply comes
in the high wailing
thunder of our
freedom–
"We do!" and in
that instant,
the we is us all,
from those whose
heads have only lifted
from their dulling labor
to see the clean white
line on blue to those
who did all they could
and yet died in their cockpits
with only the stars
their witnesses.
We are all here,
humankind, sharing
Icarus' dream.

Then the instant
is gone,
and I am, again,
in part, warrior,
predator in waiting,
weapons posed at
eighteen thousand feet,
hunting-lope at
point seven five Mach.

"Who goes there?"
Now I do not consider
the answer
but sweep the sky
ahead for
prey.

Victor Echo Zero Five

It is morning.

Arena

The sky,
soul black,
wrapped by the night,
burns with the touch
of the frozen
instantaneous lash
of storm
lightning.

Suspended in time
we wait.

Suspended in space,
we wait,
in an amphitheater
of Hell's gods.

The arena's walls
briefly reflect gray
from the white-hot
flickers
or glows from
spasms of internal fires.

We wait.

Above the arena,

Victor Echo Zero Five

bored stars stare.

No sand, but
jungle canopy
and twisted river
lay below.

We are ghosts,
gladiators' spirits,
unseen,
only heard,
as our Phantom's rumble
mingles with the storm's growls.

The night sky's dark
mingles too,
with our souls.

Last arena, lost arena,
forgotten gladiators,
we struggle to avoid trident
and flame-weaved net.

Our sword, aluminum
and fire, strikes.

We win our day's grace
and the arena floor
absorbs the vanquished's
blood.

We, victors, graced,
leave the arena,
to return to our cells.

Our weapons will be cleansed,

Victor Echo Zero Five

our turn will come again,
in the last arena.

Victor Echo Zero Five

Vietnam Epitaph II

Puzzled, he served.
Hated, he fought.
He died, and was
now
loved.

Victor Echo Zero Five

Return

Hawaii
sparkles green
on my radar
at two hundred miles.

I whisper
into olive drab oxygen mask
and Big Red
just says
"Beautiful."

Number Five,
our Number Five,
gently lunges
forward, unbidden
steel and aluminum,
eager for home,
metal feeling
the electric hunger
in my bones.

The thin white
contrail,
life line
we draw across
the Pacific sky,
fades and is lost

Victor Echo Zero Five

as we drop towards the sea.

We race in over the bay,
as we have done in the past,
but this time,
not in anger, not in vain,
not this time.

Note: About the best way to come home from a war is with a Marine fighter squadron. I was lucky to be picked for VMFA-122's return to Hawaii. so while 115 moved up to Danang, I was in the last F-4 out of Chu Lai. I put the earpiece from my portable cassette player into my oxygen mask and then, after telling the tower, "Fighter on the roll," I keyed the Play button and the Animals sang, "We gotta get out of this place." The only comment anyone made was an anonymous voice that said, as we retracted our landing gear and I turned it off, "Beautiful." The word would be said again. Big Red and I flew several legs in our bird, got thrown off Midway Island, and flew to Hawaii.

Victor Echo Zero Five

You old beast;
I scrawled "Big Hawg"
on your side
and felt boyish pride
with my name on you.

Good men, friends of mine,
also had their names
on you, and you and they
were worthy of each other.

You were there in California,
when we were strangers,
and you taught me,
an ignorant cub,
how to fight, yes,
in the sky, but also,
how to be, in the sky;

In the dark morning hours,
you and I and young John
raced in the night
near the enemy's land
and chased one home.

We returned to our patrol
and heard your battle song;

Victor Echo Zero Five

a whistling warble,
a wail of Valkyries,
and we were awed.

By day and night,
time abandoned,
we clung together, lived,
killed, in the Hoi An,
the Ashau, at Tchepone and
Mu Gia Pass, and so
many other places,
unnamed, 'til
they told me stop;
316 combat missions
was enough.

After a year, after an age,
you and I and Big Red
pointed your nose to
the morning sun
and you took us home.

I leave you
under the Hawaiian sun,
resting, engines stilled,
I stand beside you
a last time, listening
to the wind
slip over you.

It was a good thing
to know you,
Victor Echo Zero Five.

Note: VE-05 went with me from VMFA-115 to VMFA-122, where her squadron number became DC-05. I thought I'd heard the last of her then. Forty

Victor Echo Zero Five

years later, out of boredom, I fed it's bureau number, the individual number assigned by the Department of the Navy to all purchased aircraft used by either the Navy or the Marine Corps, into the web. VE-05's Bureau Number was 152990, as it was with DC-05 when she took me to Kaneohe Bay, Hawaii. Sometime later, she flew the Navy fighter squadrons VF-21 (I have a picture of her launching from the deck of USS Coral Sea) and VF-301 (I have a picture of her launching a Sparrow missile). Of course, as with her transition to VMFA-122, she does not carry the squadron number of VE-05. But that's who she was.

Victor Echo Zero Five

Homecoming

We leave the sun
and Hawaii behind.
(Strange to fly
in an aluminum bus
after life in a Fox 4.)

A handful of strangers,
a few Air Force uniforms,
a few tired wives, and
untired children.
(Strange to see
children of countrymen;
I had forgotten.)

I sit uncomfortably
in forest green
with homebound orders
on my lap.
(Strange to wear
a splash of color
under the gold wings.)

Waiting as hours crawl
at 500 knots.
(Strange to know
this is the last wait,
now homeward bound.)

Victor Echo Zero Five

Dark, now, crossing California coast,
unseen for a very long year,
and the pilot says,
"For our returning Marine,"
and dips a wing, and I see
the lights of home, spray of jewels
on San Francisco Bay, lights,
millions, on the ground, lights of home,
and I hear,
"Welcome home."
(Strange to see
lights below
after Asia's darkness.)

Lights of home.

Note: I didn't say thank you to that Air Force pilot and have always regretted not doing so. I like to think he continued doing similar things for the other returning folks his C-141 carried and I like to think he knew what his kind act meant to all of us.

Victor Echo Zero Five

SHUTDOWN
1970-1971

Note: Returning to 531 meant I was an instructor and section leader. I enjoyed the teaching but money was tight and the result was less and less flight time. Maintenance suffered and I found myself facing an in-flight emergency three out of every five flights. For some reason, it had become important to get to 600 hours in the F4, so I did not stand down until it was almost too late. With no gas and failing radios, I used our radar to get us through dense clouds and on the runway, at which point the radar also failed. I sat on the Phantom and stroked its side – every other time, that gesture had been like touching a living thing. Now all I felt was metal. It was time to go.

Victor Echo Zero Five

Yuma Deployment–1971

Five minutes after
Midnight,
Alone, again.

Southeast Asian Code
Number 114 was
"I hate this shit."

Appropriate, now,
In a border town,
Dry, windblown,
Arizona dusty,
Like
My soul.

Alone, again.

SEA Code 114.

Note: There was a long list of SEA Codes, pretty much all profane, that made the rounds at MCAS Yuma, AZ. Squadrons would go there to practice air to ground bombing on its ranges.

Victor Echo Zero Five

Pompeii's Dust

The rage of war left
Pompeii's dust on our souls.
The gray emptiness left
after the growing rot
of scientific horror
shows few signs of life,
just dull acceptance.
Trapped by our own will,
we become spiritual
suicides.

After this, what?

Is there, somehow, a yet
flickering star point of
humanity left in our hearts,
not exorcised, merely
forgotten?

Perhaps, for would we
search if it did not
live and draw us
into ourselves to look,
to understand, to overcome,
despite the many times
we died a little?

Victor Echo Zero Five

For we did, despite
the terrifying coldness,
the dead ice, of our eyes
behind green plastic;
we felt, secretly,
unknowingly,
each flaming scythe,
each tearing blast,
each diamond cluster of
death as we shook the
earth–and we shook our
souls, crumbling them,
slowly, to
Pompeii's dust.

Victor Echo Zero Five

Arizona Night

Arizona night:
stars scattered like sand.
Three miles high,
silent, watching
the incredible night.

We share the night
with none; alone
with the stars
and the desert.

Then an old orange Moon
begins to crawl
above the horizon,
yet still below
our night-racing wings.

Starfall–
our speed becomes nothing
and, still silent, we watch,
for even a star falls.

And even as it falls,
the Old One moves...
A sudden yellow glow
"Fire Warning"
and reflexes and

Victor Echo Zero Five

knowledge respond
while we turn
homeward, while
we call for help,
unseen, now,
the night beauty.

Our return is allowed,
the Old One backs away,
this night we are not
a star fall,
so later I sit in a truck,
gaudy helmet in hand,
and I look up into
Arizona night
and see again.

Dry Leaves

Something is building
within, for I feel one with
dry leaves, spun by dust devil
in time to the singing of an
unsung song;
brilliant in their dying.
(And can you hear another,
long forgotten,
Summer song?)

Fire devil, a spinning scream,
scattered us,
shattered us,
and left us.
No voices left in
burned-out throats
to sing the cold
moaning songs
of a forever
Fall.

We are others' dry leaves,
brilliant in our dying,
survivors of the bonfire,
and now scattered
on an indifferent wind,
waiting for the dissolution

Victor Echo Zero Five

of Winter,
of Spring rebirth,
which can never come.

EC-03

(d. 21 May 1971)

Echo Charley Zero Three
stopped flying
today.

It grew
tired,
old.

Two white
parachutes
and a
fireball,
and a raised
glass
were all
that
was left.

Not a great
wake,
but a
good one.

EC-03
died
quietly

Victor Echo Zero Five

and let
her men
live.

The glass
is raised.
Thank you, EC-03.

Victor Echo Zero Five

Thoughts at 43,000 Feet

Here and now,
above where others dare,
carried by whispering thunder,
I see below the darkened sea,
and wonder of flight.

Across brilliant and glowing blue,
we etch a straight white line,
which billows and fades
a quarter of a continent behind.

From here, and heart wrenchingly now,
we could mount a high crusade,
we could cause the sky to burn
and the earth to shudder,
but the wish is not there,
and our wild talents are untried.

I hear my lungs work,
breathing from deep within the fighter,
and I hear the sudden voice
from below, and then it is silent.

Whispering static is heard
while I seek out others
in this bright sky,
and the ceaseless glow

Victor Echo Zero Five

of the Phantom's far-reaching eye
shows me we are still alone.

Three feet in front of me,
silent, except for his breathing,
he looks into canopy mirror,
and behind eye hiding green visor
and black oxygen mask,
he smiles, as I do.

Nothing is said,
but we know.

We turn, curving the white line,
then losing it in descent,
to cause our own,
brilliant, sunset,
for it is already night,
seven miles below us.

Near the coast, visible in the dark,
we see the flickering rubies and emeralds
of an airliner, and briefly
I pity its passengers,
isolated in their well-lit bus;
they have not been a part of the sky.

But the thought goes on,
forgotten in the red-pool of my cockpit,
as my eyes move, seeking,
for here, in the night sky,
so near home, the danger grows.

Tired eyes may not see,
or distinguish, our running lights
from the jeweled riots of cities

Victor Echo Zero Five

and two aircraft could touch,
a waterfall of flame
and screaming metal would be all
we would leave in memory;
it has happened, it will happen.

So my eyes move, hunting,
we turn, we prepare,
the aircraft lowers landing fear,
a hawk extending talons.

Flashing strobe lights point,
we follow, and a ten thousand foot
rectangle of yellow lights says,
"This is home."
We touch the earth, gently,
for seventeen tons, and I watch the lights;
their passing slows, after a mile,
and we follow a path,
bordered by steady blue diamonds.

In the dark I see flashlights,
up thrust by t-shirted Marine,
who, chewing his gum impatiently, awaits us.

Obediently, almost cowed, we follow his signals
until the fighter is in line
with the already bedded down others.

A pause, a voice,
"Ready for shutdown."
and the whispering roar dies.

I undress myself of the aircraft and
emerge, descending in the darkness,
and we walk around her, checking,

Victor Echo Zero Five

silence broken by clicks of
cooling steel
and we turn away.

Stiff and encumbered, out of our element,
we walk slowly towards the hangar lights,
and I look up to see the stars, to say,
"Good flight."
and he agrees, briefly, and in silence still
we both wonder how to tell the other
we mean the aircraft, for with her
there had been three of us
in that bright sky.

Victor Echo Zero Five

Vietnam Iliad

Anger be now your song,
for the gentle ballads of youth
were hushed in the roar of fire,
soul-eating flames,
leaving bitter ashes.

Your rage sings
like a whistling sword,
cleaving the tenuous hold
others made on you.

You do not feel the gentle wind
of home–this soil
does not hold the print
of your foot.

What war do you now fight?
Who is your enemy,
when you face you?
Stilling the guns
only stops the killing;
the dying continues
as you destroy you.

Victor Echo Zero Five

Mount Out

They awakened to the dull buzz,
which displaced the heavy, wet night.
Yellow light pooled; bedroom, bathroom, kitchen,
but no dawn yet.
In the morning cool, dark, they kissed,
and she watched him leave.
Her house was still and calm,
his squadron was flexing and stretching
in the Carolina morning.
The babe awoke, cried,
while great engines stirred and whined.
She lifted the child and caressed him,
as great, gray shapes taxied out,
with the sun streaking fire in the east.
She gave the child her breast,
felt its hungry, powerful, thrilling pull,
and her man caused dragon's thunder;
a mile away, but the howling roar came,
and she caressed her child's cheek
feeling her man's thunder,
vibrations in her child's flesh.

Victor Echo Zero Five

Investigation

The night sky
sees falling fire
touch a mountain's flank,
and a burning rose blossoms
and fades.

Men come to probe
through skeletal metal,
shattered and burned,
scattered through forest,
shocked into silence,
under hushed sky.
Graphs and figures,
computations and evaluations,
documented is
How.

But even after the land heals,
after the forest slips back
to surround, embrace,
the last few metal shards,
even after the names
of the fallen are forgotten,
never is found the
Why.

Victor Echo Zero Five

Tribute: Last Flight

I

High in winter's sky,
we watch Sun fall and Moon rise.
Chased by the black,
red flames briefly in the west,
pinkly caresses clouds and sea,
then is gone,
leaving only a fading glow,
soon lost to a splash of stars.

II

Stars;
each a brilliant, distant, beacon,
serving as challenge
and monument.

III

The stars called to us,
to all who travelled
through those endless caverns
and ghost mountains of wispy
clouds, called and challenged.
To be a part of the sky
we dreamed and strove

Victor Echo Zero Five

and in the vanished paths
of others, now gone,
we made our way
and learned of the sky's glory.

IV

Some fell,
the Earth received them.
Icarus dared too much,
but was remembered;
some were not, not by men,
and I see faces gone,
lost chasing the wind,
their names not long noted,
but in winter's sky,
in darkened cockpit,
I look into the eternal sky
and see each one's monument:
a star.

Victor Echo Zero Five

OF A FIRE IN THE NIGHT
1971-1990

Victor Echo Zero Five

The Wasted

It was us;
seek out
engage,
destroy,
it was us,
in harm's way,
and the gray
monsoons obscured
the vision
of eye and
mind.

We lay here
in accordance
with their
will and
confusion.

In our
Flanders' Fields
no poppies,
but rice;
the mud and
the death
were the same.
Wasted;
gladiators, pillow-armed,

Victor Echo Zero Five

blinded,
crippled,
reviled
by the mob,
loved only
when we fell.

Wasted,
returning
bitter and
armed, forgotten
centurions
who may turn
on their Rome
and leave it
wasted.

Victor Echo Zero Five

Long Gone Time

Leaves
friends
with uncaring wind
gone
forgotten
in the glory of new buds.

Fences, walls
some remain
some change
past time sits
and mocks
on your shoulder
as you walk on
unknown and familiar
paths.

Echoes
of the past
long gone
faded, worse,
forgotten
and by slight magic
of terrible time
the unfamiliar
the stranger in this land
is the one who comes this way

Victor Echo Zero Five

again
as I do.

Victor Echo Zero Five

27 Jan 73

They sat
at a round green table
as large as the
target bull
on the Arizona sand
where we had practiced
our wild and deadly
trade.

They signed paper;
paper covers rock,
rock breaks scissors,
the game is ended.

The incredulous dead
lay in wonderment
at the miracle.

The war has ended.

How many years did
I contribute
in thirteen months?

Blood and mud,
sky and smoke,
fire and pain.

Victor Echo Zero Five

What have I received?
What have I given?

And why?

Gray steel and aluminum
shield and sword
warriors' insignia
burned in foreign
skies
as did
my soul.

Where am I now?

The flame that touched my heart
when I saw the great warbirds
challenging the sun
begins to cool;
the guilt
at being alive and out of the jaws
of hunting technology
may now be eased, grown past,
perhaps now I can return,
and the end be a
beginning.

Victor Echo Zero Five

For Terry Graves

Peace, the man said.
Sorry, Terry,
Too late
to halt your eagerness.

You died,
and people laid
claims to the value
of your death.

They cannot say
for Nothing or
for Glory or
for Duty;
each man's
death is his only
private understanding.

You placed yourself
in the path of flame;
for your reasons
you teased fate
and pulled the
Old One's beard.

And one time
you lost.

Victor Echo Zero Five

Only you can say,
now, if the price
was worthwhile.

I still wish,
Terry, you were
here.

Victor Echo Zero Five

Held Fire

Clenched
in soul's grip,
latent coals
sparked with
fierce breath,
blazed,
wolf eyes
glinted moonlight.

The time
is changed,
the place
is changed,
the fire
remains,
patient,
waiting,
soul held
fire.

Victor Echo Zero Five

Those Long Time Rememberin' Homecomin' Memory Blues

Oh, lady mine, c'mon and listen now,
while I lay my story down, far down,
low down, as low as I feel down,
rememberin' my homecomin' memory blues.

We shook that sand, yeah, that Chu Lai sand,
shook that sand from our boots, one cool mornin',
a while ago, me and Red, a while ago,
and we climbed aboard old Number Five,
my Number Five, Old Hawg, big Hawg,
she sang a whistling thunder song,
song bird, war bird, big movin' bird,
and we turned her nose east,
and surprised her, aimed her homeward,
took her homeward, she took us homeward,
but now I got those long time
rememberin' homecomin' memory blues.

We came arcing, over the sea,
came rumblin' and hurryin',
across a Hawaiian bay, home bay,
and echoed our lovin' thunder
off old lava rocks
and let the people know,
we were home, baby!

Victor Echo Zero Five

I remember her, where we left her,
under Hawaiian sun, alone and quiet,
chained to the earth,
and the sadness comes,
Big Hawg, song bird,
causes me to remember,
lookin' over my shoulder,
what I can't seem to lose are,
those long time rememberin'
homecomin' memory blues.

Old Number Five, good Number Five,
causes me to remember, sweet Number Five,
and sweet memories come of Number Five,
of chasin' clouds and wind and stars,
and diggin' it and us and her,
and lovin' it and us and her,
the sweet buries the sour,
and first remembered are the good,
yeah, the good ol' boys who came by me,
Big Red, and the Goose, and Skip,
and Johnny Fogg, and Mac,
Sugarbear, and Johnny Cool,
and a bunch of good ol' boys,
sittin' and grinnin' in the Chu Lai sun,
with tired and squinty eyes,
good memories, good time, lost time,
forgotten is the war time,
the agony time,
the killing time,
the why those eyes were tired and squinty time,
but dig it, the good time, the movin' time,
the blazin' blue forever time,
and it's a gone time, a lost time,
so I got those long time

Victor Echo Zero Five

rememberin' homecomin' memory blues.

Victor Echo Zero Five

On the Fall of Pleiku

I

The sound of dragon, laughing,
slapped across South China Sea.
Gray scud swirled,
monsoon rain drifted by,
and we who lay here
in accordance with your will
speak not.
Rain falls yet,
rice grows still,
blood falls yet,
steel and flesh,
both living,
shuddered on Asian earth.

II

Echoes fade, as do we,
specters best unseen,
dark shapes, now, with dim eyes,
flecked by burning red,
mottled green on arms,
thin, haggard, young, dead,
but the eyes! See our eyes!
Ours was ever not to ask,
and we did not,

Victor Echo Zero Five

but did all we could,
and you know us not.
We lost much in the clinging mud,
we suffered much in the torn sky,
but you lost more,
for you do not know
our names
any more.

III.

Remember how we were?
We swept past, raising dust,
our clenched hands aloft,
grins slashing white
beneath helmets
faded tan and green;
our joy our youth,
for the pain was a century
away–
it was away, it was
tomorrow.

IV

Came the fire
and the dark
and the long cry forever
fall.

V

In our pain and rage and death
we lashed out, fought,
as we were expected to,
until the earth wept.

Victor Echo Zero Five

You watched,
a new game,
on cool screens
of glowing electrons,
and blended our body
count
with the violence
of entertainment
until none knew
the difference;
but we did.

VI

It was expedient
to spend us.
It was expedient
to forget us.
Principle and reason,
justification and purpose,
washed off the pages
the news stars read,
washed and blotted
by a trickle
and a flood
of our tears,
blood,
sweat,
and spit.
What was left of us,
after you made us
numbers?

VII

It is sour rice,

Victor Echo Zero Five

but rice still.
You do not have
heart enough
to taste it.
The years are dust
and you shake them
from your feet
as you turn away,
and the dust drifts,
remembers,
settles on our
last formations,
the neat, endless rows
of white on green,
an order in death
never present in
life.

VIII

Our question is
your curse:
"Why?"

Victor Echo Zero Five

Last Night

Last night the wind laughed,
sparkled in south Ohio ravines,
played with my shaggy hair,
hugged me with warm breath.

Last night
peace whispered through the trees,
home's fragrance caressed about me,
moon light fell warm on my land
as I walked from the creek.

Last night
the lights of my house beckoned,
sanctuary was all about,
I turned, I saw, I was safe,
until I looked upward.

Last night
I saw again the indifferent stars
and recalled against my will
those same stars who coldly stared
as I tumble across them.

Last night
was another and a dreaded night,
when blood colored white stars red,
and sky's thunder was blood's thunder

Victor Echo Zero Five

was war's thunder.

Last night
a burning roaring remembered horror
was split by my pup's nuzzlings
and now old sweat dried
as gentle wind brought me home
last night.

Victor Echo Zero Five

Late Evening Questions

"What was it like?" she asked,
"I must understand," she said,
"My man was there
and I must understand.
What was it like?" she asked.

I sat, watching dust motes, silent.
"What was it like?"
Fire blossoms recalled;
The earth torn,
the sky dead.
"What was it like?"
The long fall forever,
A generation of dry leaves,
A taste of bitter rice.
"What was it like?"
Drudgery and pain,
Hunter's elation,
Prey's terror.
"What was it like?"
Medusan beauty,
Combat's hours traded
for years.

"What was it like?"
Our bombs cracked their bunker,
Our napalm ate their air,

Victor Echo Zero Five

And they died,
That time.

She sat, watching my silence,
Waiting for the coherent words,
The structure,
The bridge,
And I was silent.

"It depends," I said,
"It depends."

Victor Echo Zero Five

Secrets

They listen to the words,
hearing them how they need to hear them,
stories, subplots, of a war.

I tell them nothing
while reciting history of incidents–
hidden, guarded, I keep myself
from their too sheltered glance.

They hear the story of howling thunder
echoing through Laotian hills
as flame smears earth and sky.

A warrior's story excites, seduces,
these protected and safe ones,
who lean forward, entranced
by Kali's dance.

Unknowingly in their shuddering
fascination with the swaying cobra
of combat, they are glimpsing
the warrior's only secret,
what I do not tell them,
what is all that is worth telling
about war.

I loved it.

Victor Echo Zero Five

For minutes or days,
out of madness or exhaustion,
all rationalizations come to mind,
as I feel back to that
phantom hunter
and know again the burning,
joyous rage, that cold fire
of heart.

The addiction, the seduction,
was complete.

Now here, I talk about the war,
about combat, about everything,
about nothing,
for I will not share
the warrior's love with them.

They would claim incomprehension
in self-defense, for their
half-held breaths and hungry eyes
show their attraction to the dance,
and this they would never admit.
(The secret is I can see me
 in them but they cannot see
 them in me.)

Victor Echo Zero Five

Brothers and Sisters

Introduction:

It was the fire,
last time, last night,
the dragon, slain,
had left its teeth
in my heart,
and dragon's teeth
...are seeds.

Metallic irony–
to survive is
to guarantee
the struggle's life,
but not the
survivor's.

Not a day
without the taste
of bitter rice.
Flashes of light
come unbidden to
the hidden dark
and that which was,
is;
home from the hill,
I am,

Victor Echo Zero Five

but the hunt
lingers.

What songs now?
What instruments
could play them?
I hear a symphony
of screams,
thudding of the guns,
roar of the flames,
wrenched bits of voices,
all held in terrible
cadence by cold
silence.

I

The Ohio runs deep,
etched wide in the dark earth,
like the bloodline
of brothers.
Two brothers, sharing
Ohio bloodline,
two wore splattered green,
and sought that place
where,
"Here, there be dragons."

He went among the tall grasses
where more than the weeds
tore flesh.
Young man, squinting in
the sun,
tasting the balance of life
in the sweat on his lip,
carrying steel and plastic

Victor Echo Zero Five

of weapons
with a curious grace,
a hunter's ease,
a prey's caution,
while feeling time
ooze past.
(Sudden streaks of fire
and feeling,
as death and hate and fear
burst out, then
fade.)

I went another way,
tumbling across the sky;
the stage was different,
the play was the same.
The fire in the sky
was the fire on the earth
was the fire in my heart.
A Fury possessed me,
drove me past fear
to an ice hard rage.
(And I did not remember
ice melts.)
The dust across Asian sun,
raised by my passing,
settled on
my soul,
as eyes grew dull
behind green visor.

II

Johnny didn't come marching;
the parade was called off–
no rain checks were issued.

Victor Echo Zero Five

But it rained.
None of us were outlaws;
the image, too romantic,
never formed,
even when the tread
of our tired boots
echoed in silence
on the streets.
Lepers were at least
cast out;
we were, terribly,
ignored.
("I do not believe,"
the young man with the
scar
said, "we were forgotten;
we were too hated
for that.")

III

Like the Ohio
we've rolled on,
rolled on–
the river, at the
same point,
is changed,
changing,
never the same.
What do I see
in my brother's eyes–
is that flicker
rage's ember,
left from the
burning elephant grass?
Or is it,

Victor Echo Zero Five

just,
the reflection of my own eye?

Our brothers and sisters are
many, now,
a bloodline
broader than the
Ohio;
from here to the
Mekong–
our family is larger
and lonely.

IV

Trees grow,
even in the
dark night,
as do families;
a brother of our
fire family
shares a smile,
a sister of our
fire family
shares a word.

Gentle touch is made,
the tree flowers,
even in the dark,
with warm, glowing buds,
which, rushing,
blossom, and the
dark fades into
the earth
as we cling
to one another.

Victor Echo Zero Five

Warm light
of brothers and sisters
finding one another
melts the soul ice
and brings
hope.

Victor Echo Zero Five

Of a Fire in the Night

Blue-black the night, broken
by darker shadows, obscured stars
hidden by silent scud, cold wind blown.

Borne aloft, chained to swift
whistling thunder, yet silent,
hearing only electron's hiss
in flight helmets, eyes flicking
between the outside dark and dull
red glow of instruments and scopes.

Yellow and red fire smear obscene
flowers across the earth, hunting
tracers stab upward and we are drawn
to their flashing source as moths
to the flame.

A two mile stoop, past the vertical,
driven faster than any hawk dared
dream, driven by the banshee screams
of fire-tailed engines, driven
by our hunter's madness, we
fall,
the long fall forever.

A decade later, the fire in the night
is still now, still felt is the mask

Victor Echo Zero Five

and helmet, still heard the controlled
voice,
 "Ajax in on the guns,"
and still felt is the surge of
hunting feeling, and fear, and
anger, and hate,
all felt again, each night,
an eternal fire in the night,
traces left by dragon's passage.

Don't tell me of your fear of
the dark, and what it holds,
for I have ridden lightning,
and still find comfort in its
heat and familiarity.

Victor Echo Zero Five

Rain

Cool San Diego rain
drifts in from the beach,
silent in the dark night
and lightly touches me
as I walk to the motel bar.
I stand in the dim light,
letting tiger-stripe shirt dry
and tired eyes search for a friend.
"Is it raining out?" he asks;
 (Vietnam, Class of '68, 101st)
"Off and on," I reply;
 (Vietnam, Class of '70, VMFA-115)
"How hard?" she asks,
 (Vietnam, Class of '67, 93rd Evac)
"Early monsoon," I say,
and they all nod.

Chance remark brings it back
and there is a silence,
only a few seconds,
within which are days,
weeks and months,
and then we shift, we are
here again,
in San Diego,
attending a conference.

Victor Echo Zero Five

Later, rain stops,
California sky clears,
chamber of commerce stars
and moon shine across
slow ebbing Pacific.
But somewhere inside
there is always
early monsoon rain.

Victor Echo Zero Five

Charley/One/Four Remembered

Black wall,
warm to the tentative
touch,
as I caress a name;
Terry, Dick, or
Gentle Ben–
friends frozen in time
of memory
despite September's sun;
the memories, now bidden,
come as I see names and
myself
in polished black stone.

Here is Terry,
shot out of the sky,
Dick, vanished in the night,
Ben, bucked the odds too often,
and others, grouped as they fell,
so memories come:
Most of a squad of
Charley/One/Four–
I'm sorry,
I'm sorry;
we came as fast
as we could,
but it was not fast enough

Victor Echo Zero Five

that awful monsoon-smeared day
and even as our thunder
echoed up the Hoi An valley,
even as you heard us coming,
you died.
We caused the hills
to burn
even in the rain,
so some lived.

But what is remembered
is the group of names
of December, 1969,
warm to the touch
on a black wall.

Victor Echo Zero Five

Frags

We sat up 'til two,
strangers, yet brothers
and sisters of a
family we didn't know
was being forged by
fire and rain,
by pain, and
by abandonment.

The talk slid about,
touching this and that,
tentatively, then settled,
gently, on the fragments
making our family.

She'd been a nurse,
he was in tracks,
Dave was oh-three-oh-shit,
Mike was among the first,
another was among the last,
Phil was in the Fourth,
someone was doorgunner,
I flew in Phantoms,
and the places were familiar,
were very familiar, even when
we hadn't been there–
the Mekong, the Z, Rockpile,

Victor Echo Zero Five

Hue, Dak To, Ia Drang,
Ashau, Tchepone, Parrot's Beak,
Long Binh, Ban Karai, Thud Ridge,
–places always seem familiar
when a family member's been there.

Hours later we depart,
to Oakland and Las Vegas,
and Seattle and Anchorage
and Philadelphia
and once again we become
frags.

Victor Echo Zero Five

Interview–Vietnam Veteran Memorial, 1983

"What do you think,"
she asked, microphone
discreetly held low,
"people should know about
the Vietnam veteran?"

I answer, reason,
compassion, and healing
my tone,
and we are all
the same,
we are all
brothers and sisters,
we all need
to heal together,
whether or not
we were in Vietnam.
I say all this,
as I have a
thousand times.
My voice is one of
reason and compassion
and healing,
and the words are all
true.

Victor Echo Zero Five

But my heart rages
silently,
and screams in protest,
silently:
We are not the same—
you sent us, you let us go,
you equivocated while we fought,
you tied our hands while we bled,
you abandoned us while we died,
you disowned us while we returned,
and in front of this black wall
of remembrance of my true
brothers and sisters,
you dare ask me that question,
you demand again that mine be
the voice of reason, compassion, healing.

As you hear my voice, do you even guess
at the heat of my
long-held fire of rage
and the shriek I'd rather speak?

Victor Echo Zero Five

On Distance

Wednesday afternoon, District of Columbia,
Scud running low in November sky,
And I stand before polished black granite,
Seeing myself reflected through carved names,
And through unbidden tears of memory.

I reach out to touch the wall and a name,
And a forgotten memory, and a part of myself,
And the stone feels warm, like the handclasp
Of a friend.

Beyond the treeline, brooding in shadow,
Lincoln sits in frozen judgment,
And the path from there to here is worn smooth
By my brothers and sisters finding their way.

Later comes time to lean against a tree,
Feeling the slow pull of gentle Fall breeze,
And think of the path from Lincoln to us,
And think of deep running roots.

He might have understood our pain,
Having seen our nation in the agony of division
Once before, families split, war's awful,
Awesome, cost and confusion.

He knew Vicksburg, and Seminary Ridge,

Victor Echo Zero Five

And Fredericksburg and Petersburg,
And the Wilderness, and the hundreds of stones,
In hundreds of towns, carved with names.

The path between the monuments is not long,
Short as the time in a nation's life
Between Shiloh and Khe Sanh, so much the same,
Despite the differences of place.

The important parts were the same, a century apart,
Courage, and pain, and death, and the color
Of American blood was the same and more important,
Than blue or gray or mottled camouflage.

Lincoln would have understood sacrifice,
Even for goals never attained, out of reach,
And he would have honored the offering made,
Unrequited gallantry is not bound by time,
And we who are joined across the century
Did all that was asked, and more.

There is value, Abraham knew, in the offering,
And if some cannot honor the giving,
They dishonor themselves.

Shiloh, Thud Ridge, Gettysburg, Khe Sanh,
And black granite in Constitution Garden,
So close together,
Under Lincoln's gaze.

Victor Echo Zero Five

Serenade Echoes

She loaned me a book,
  thought I might enjoy it.
A book about flying, B-17s,
  World War Two and ancient history.
A serenade to a big bird,
  and I read words from forty years past
and found I heard echoes,
  blurred together,
of Fortress and Phantom.
  He wrote of the skies of Europe;
mine were Asia's.
  And he wrote of friends down
and wanting things better,
  of ugly flak flowers
and quiet moments talking
  with a friend.
And he wrote of the tears
  which rise up from the depths
where we bury them.
  The abyss of grief.

I wanted to find him, and talk,
  and ask how things are going,
and maybe trace patterns in the air
  with hands, talking flying,
and tell him about Mu Gia Pass flak,
  and friends down,

Victor Echo Zero Five

and the tears
  and ask him
if it gets better
  after forty years.

I hurried into the book,
  skipped into the first chapter,
left pages folded over on one another,
  which waited in patient silence
like an untripped boobytrap,
  and did not see
the "Note on the Author"
  from an anonymous Britain
that after thirty-five missions,
  when he could have gone home,
he volunteered for Mustangs,
  and died over Hanover
26 November 1944.

It was, again, a friend down;
  Jack, or Peck, or Ben,
or Bob, or Mike, or Joe,
  a friend down,
and the same thoughts came
  four decades late
of wanting to have been there,
  maybe I could have covered him,
maybe I would have spotted
  the guns or the SAM or the MiG.
But he was not lost to SAMs
  or MiGs; wrong war.
And I could not have been there,
  when there was before I
was born.
  Still...

Victor Echo Zero Five

He hauled bombs to Berlin,
  and to Hanover, and Normandy,
and hoped it would make things
  better.
And maybe all the kids
  who climbed into Fortresses,
and Focke-Wulfs,
  and Lancasters,
and Bettys, and Mirages,
  and Phantoms, and, yes,
and MiGs
  have all hoped
it would make things better.
  He never found out,
26 November 1944;
  I still wait to find out,
14 July 1989.

He felt the stream of the sky
  cover the world
and saw divisions were
  imaginary.
I recalled a morning sun
  spreading its light beneath me,
sweeping across borderlines
  invisible in the light;
existing only in what might be
  a special dark,
and I had put away the thought,
  as did the kids in the Fortresses,
in Focke-Wulfs, Lancasters,
  Bettys, Mirages,
and, yes, as did the kids in the MiGs,
  so I could pretend
the lines were real.
  Why haul the bombs

Victor Echo Zero Five

if the lines were not
  real?

I have seen the films
  a thousand times,
gunsight flashes
  and flickering images
of planes falling,
  sometimes burning,
sometimes not,
  sometimes whole,
sometimes not,
  falling to earth
or sea, or hanging
  in the air
while cannon rounds
  kick chunks away
in slow motion.
  I see the films
and each time
  I look for the white blossom,
a 'chute that seldom comes,
  and I don't care
that the plane is theirs
  or ours.
They are all mine,
  in my heart,
and I want them all
  to make it back
to the world.
  And they cannot,
their forever fall
  all that remains preserved,
shadows and light
  now entertainment.

Victor Echo Zero Five

She said she didn't mean
  for the book to hurt,
but it was the hurt I know
  and would never give up
for it is the cost
  of the joy, the love,
the warmth of more than
  that of the tears.
He and I never met,
  never talked,
and we flew together;
  we shared the sky
and the love of clear blue
  and midnight stars
and machines come alive,
  and the absolute honesty
of flight,
  tracing contrails,
life lines,
  in arcs joining
science and passion.
  And in the moving wind,
which respects no border,
  faintly comes to me
echoes
  of a serenade
to a big bird.

–For Bert Stiles

www.ingramcontent.com/pod-product-compliance
Lightning Source LLC
Chambersburg PA
CBHW061326040426
42444CB00011B/2788